Table of Contents

Disclaimer

This Handbook includes links to documents and information on non-EPA sites. Links to non-EPA sites and documents do not imply any official EPA endorsement of, or responsibility for, the opinions, ideas, data or products presented at those locations, or guarantee the validity of the information provided. Links to non-EPA web sites and documents are provided solely as a pointer to information on topics related to area contingency planning that may be useful to EPA staff and the public.

While EPA will attempt to keep links to information timely and accurate, the Agency makes no expressed or implied guarantees. EPA expects to review this Handbook routinely and update the links listed in the appendices as necessary.

This page intentionally left blank

Introduction

This Handbook is a guide and reference for the development of Area Contingency Plans (ACPs) for environmental emergencies. While it is primarily intended for use by EPA emergency response program personnel, area contingency planning is necessarily an inter-agency process, and the use of this handbook to inform other agencies of EPA's planning process is encouraged. Because area plans are focused on specific geographic domains, with many physical and jurisdictional variables, there can be no 'one size fits all' plan format, but maintaining a national consistency in the basic content is important, particularly considering the statutory and regulatory requirements by which EPA and other agencies are bound.

This handbook was developed by EPA's Area Planning Workgroup during 2011 and 2012 and incorporates the accumulated knowledge of years of contingency planning experience. Although ACPs are specifically mandated by the Oil Pollution Act of 1990 (OPA 90), EPA's responsibilities under other laws, including CERCLA, make an all-hazards approach to contingency planning desirable. The processes of planning for responses to all types of environmental emergencies (e.g., oil spills, hazardous materials releases, natural disasters) share common elements that have been demonstrably successful in major responses.

In the interests of conciseness and accessibility, this handbook will not recapitulate extensive portions of related documents, but will list key references, including laws, regulations and technical resources, in appendices.

This Handbook is available for download as a PDF file from EPA's Office of Emergency Management web site at *http://www.epa.gov/oem/*.

This page intentionally left blank

Overview of Area Planning

A. What is an Area Contingency Plan?

An ACP is a reference document prepared for the use of all agencies engaged in responding to environmental emergencies in a defined geographic area. Throughout this Handbook, the terms 'Area Contingency Plan' and 'ACP' also encompass the processes for developing and managing Sub-Area Plans and Geographic Response Plans, which have scopes more limited than the ACP itself.

Under federal law (OPA 90) and regulation (National Contingency Plan), all United States territory is divided into jurisdictional zones, with the U.S. Coast Guard (USCG) designated the lead agency for planning and response in coastal zones and certain major inland water bodies, and EPA designated the lead for inland areas, with certain exceptions for DoD-managed areas. As an EPA document, this handbook is focused on inland zone planning, but EPA also has a role in the coastal zone, particularly regarding oil spill counter-measure concurrence and approvals. EPA-lead inland plans covering areas adjacent to the coastal zone must also ensure compatibility with USCG-lead plans for those zones. Appendix A provides specific details on federal jurisdictions.

Under CWA 311(j)(4), there are specific required elements for ACPs. These elements include:

- ❑ A description of the area covered by the plan, including areas of special economic or environmental importance that might be damaged by a discharge. This description should provide a comprehensive picture of the defined area, which may be a body of water, a watershed or a political jurisdiction

- ❑ A description of the responsibilities of owners, operators and federal, state and local agencies in responding to a discharge, or mitigating or preventing a substantial threat of discharge. The plan should identify those entities with authorities and resources for planning and response, describe their capabilities and establish an operational framework for these entities to ensure optimum communication and coordination during a response.

- ❑ A list of resources (personnel, equipment and supplies) available for response to discharges.

- ❑ A description of procedures for expediting decisions on the use of dispersants.

- ❑ A description of how the plan is integrated with other plans.

When implemented in conjunction with the NCP, the ACP must be adequate to remove a worst case discharge, and to mitigate or prevent a substantial threat of such discharge. Additionally, ACPs may provide guidelines for conducting specific tasks such as: sampling, classifying, segregation, and temporary staging of recovered waste; and identifying prior state disposal approval, various waste disposal options and a hierarchy of preferences for disposal alternatives (40 CFR 300.310(c)).

An ACP is not a rigid, prescriptive plan with step-by-step instructions for responses. Rather it is a mechanism to ensure that all responders have access to essential area-specific information and to promote inter-agency coordination as a means of improving the effectiveness of responses.

B. How is an ACP developed?

An ACP is the product of a collaborative process involving stakeholders within the defined area, organized as an Area Committee (AC). Although it must be initiated and led by a federal agency (33 U.S.C. 1321(j)(4)(B)), the AC is to be comprised of members from qualified personnel of Federal, State, and local agencies. The AC provides a forum for agencies to develop constructive working relationships while identifying issues and problems and developing solutions in advance of a response. The AC is responsible for developing the ACP, evaluating its implementation, and maintaining it through a continuous improvement process, through consultations with the RRTs and others as appropriate.

C. What are the benefits of an ACP?

Responding to the immediate circumstances of an environmental emergency can be a challenging task. Overlapping jurisdictions and potentially divergent interests of the parties involved can make response even more difficult. The ACP provides a mechanism for planning for these potential complications prior to an incident. The ACP is a useful tool for responders, providing them with practical and accessible information to place the incident in a larger context, informing them about who and what they need to know to make the response effective.

The process for ACP development may be as beneficial as the final product. The Area Committee provides a forum for all parties to identify problems, resolve conflicts and become informed about the issues raised by actual and potential incidents. From EPA's point of view, the AC provides an effective mechanism for informing a wide audience about the response and planning concepts as part of the National Response System (NRS). The NRS is the government's mechanism for emergency response to discharges of oil and the release of chemicals into the navigable waters or environment within the jurisdiction of the United States. The NRS functions through a network of interagency and inter-government relationships that were formally established and described in the National Oil and Hazardous Substances Pollution Contingency Plan (NCP) as found in 40 CFR Part 300. The AC provides a way for local, state, and federal members to define their most significant concerns, ensuring that they will be considered should a response be required.

D. What are the statutory and regulatory underpinnings of the ACP?

ACPs were initially conceived in the context of oil spill responses, but the ACP concept has grown beyond that to encompass the prospect of responses to environmental emergencies in general, including hazardous materials releases, natural disasters and acts of terrorism. There is a substantial foundation of laws, regulations and executive orders that provide the basis for ACPs. These include:

Clean Water Act (1972): The CWA (originally the Federal Water Pollution Control Act) expanded the federal government's authority to regulate discharges to waterways and provided the original statutory basis for the National Contingency Plan (NCP).

Oil Pollution Act of 1990: The OPA 90 amendment to the CWA established ACP requirements for the NRS to address worst-case discharges of oil and hazardous substances and mandated facility-specific plans (facility response plans [FRPs]) for certain categories of facilities.

Comprehensive Environmental Response, Compensation, and Liability Act (1980): CERCLA established a federal emergency response program to deal with immediate threats from hazardous substances and pollutants (excluding petroleum as provided by 42 U.S.C. 9601(14) and (33)) and a remedial response program to deal with hazardous waste sites requiring actions consistent with a permanent remedy

Emergency Planning and Community Right-to-Know Act (1986): EPCRA amended CERCLA by adding requirements for community-based emergency planning, through State Emergency Response Commissions (SERCs), Local Emergency Planning Committees (LEPCs), and public disclosure of hazards associated with certain facilities.

The Robert T. Stafford Disaster Relief and Emergency Assistance Act, as amended: The Stafford Act provides the authorities and funding for federal support to state and local entities in responding to major disasters and emergencies.

National Response Framework (2008): The NRF is the federal executive document that provides the national blueprint for how the Nation conducts all-hazards response.

National Contingency Plan (40 CFR Part 300, amended in 1994): The NCP is the federal regulation that defines the authorities and responsibilities of designated federal agencies for responding to releases of oil, pollutants and hazardous substances.

Regional Contingency Plans: The NCP requires that each federal region, through its Regional Response Team (RRT), develop RCPs. ACPs exist under the umbrella of the RCP.

Homeland Security Presidential Directives: HSPDs are executive orders that address specific issues. HSPD-5 covers incident management, and requires the establishment of the National Incident Management System (NIMS). HSPD-7 addresses the protection of the nation's critical infrastructure. Presidential Policy Directive/PPD-8 focuses on improving the overall preparedness of the nation to respond to emergencies. PPD-8 replaces HSPD-8.

State Laws: Each state, territorial and tribal entity has its own laws and regulations that apply to environmental emergencies. As partners in the ACP process, these entities identify which agencies and requirements are relevant to the ACP.

Local Laws: Each locality participating in the ACP identifies which of its laws and ordinances are relevant to the ACP and which agencies will participate in the ACP process.

A more detailed summary of the statutory and regulatory basis for ACPs is included in Appendix A.

E. What is the relationship of the ACP to other plans?

The NCP is the regulatory foundation for interagency contingency planning. The NCP is a regulation that establishes the authorities, responsibilities and relationships of agencies in responding to environmental emergencies. RCPs extend the NCP model to a narrower regional focus, bringing in states and other entities as participants to deal with region-specific concerns.

In Executive Order 12777, the President delegated the authority to designate areas, appoint Area Committee members, determine the information to be included in ACPs, and review and approve plans for the inland zone to the EPA Administrator. The EPA Administrator, through delegation 2-91, initially designated thirteen geographic areas already covered by Regional Response Teams, and the Regional Response Teams as the initial Area Committees. The EPA Administrator also delegated Regional Administrators the authority to designate different geographic areas within their Regions and appoint different Area Committee members. Regional Administrators are authorized to delegate the authority no lower than the Division Director level. For this reason, an RCP may function as an ACP, if the designee determines that there is no need for formally defining multiple ACPs within a region. Sub-regional concerns may also be addressed by Sub-Area Plans, which have a more limited scope, but many of the same elements as ACPs or by Geographic Response Plans, which focus on specific response strategies and tactics for more narrowly-defined areas. If the designee determines that the RCP will serve as the sole ACP for the region, the RRT assumes the responsibilities of the AC as described in 40 CFR 300.205(c). In this case, the RRT solicits states for local representatives. NGO and private sector members may also be invited.

ACPs also interface with plans developed by state and local authorities and by private sector facilities, as well as other ACPs in bordering jurisdictions, such as those developed by the USCG. The diagram on the next page illustrates the relationships between the various plans. The Federal Response Plan (FRP) was superseded by the National Response Framework (NRF) in 2008.

Relationship of Plans

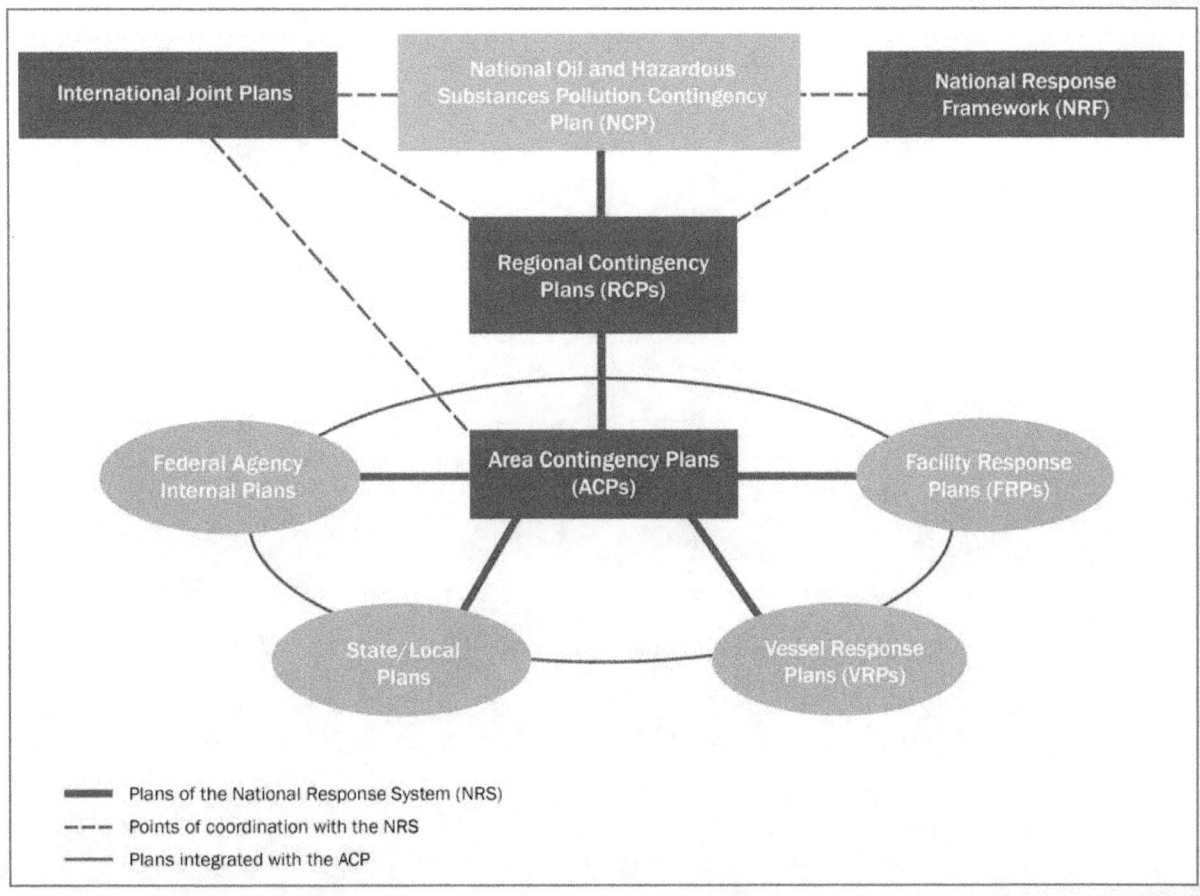

There are three levels of contingency plans under the national response system: National Contingency Plan, Regional Contingency Plans, and Area Contingency Plans. The relationships between these plans and other planning mechanisms are described above.

This page intentionally left blank

Initial Steps/Preliminary Analysis

The decision to initiate the development of an ACP (or a Sub-area Plan or a Geographic Response Plan) should be preceded by an analysis of the costs and benefits of the process by the federal lead agency's designated On Scene Coordinator. The initial forum for this analysis is most often the RRT, because ACPs are subsets of the RCPs. The stimulus for the development of an ACP may be the experiences of agencies during the response to a major incident, or it may be a pro-active effort to protect sensitive resources or to address issues related to high-risk facilities. There are no constraints on the ACP-defined areas within the region. It may be based on jurisdictional boundaries, for example, if a state perceives the need for specific measures within its boundaries, or it may be based on geographical determinants, such as a watershed that encompasses sensitive resources. Regardless, if there is a consensus among the RRT participants that an ACP should be considered, then the RRT should establish an ad hoc committee of interested agencies to conduct an initial analysis, which should involve the following considerations:

1. An inventory and assessment of existing plans, including the RCP and any other federal, state, regional and local plans, and an assessment of the effectiveness of these plans, including the identification of gaps and other inadequacies that an ACP could remedy.

2. Identification of the portions of existing plans which are adequate and can be incorporated into an ACP.

3. Identification of potential sub-areas within the ACP boundaries that may require special attention, leading to sub-area plans.

4. Review of data and information from past incidents (e.g., after-action reports, lessons learned, unresolved issues). This review should identify specific problems that the ACP will address.

5. Preliminary identification of sensitive areas, including environmental, cultural and economic resources.

6. Identification of actual and potential jurisdictional conflicts.

7. Identification of high-risk facilities and critical infrastructures.

8. Assessment of natural disaster risk and impact.

9. Initial estimates of the time and resources required for developing the ACP.

10. Preliminary identification of key agencies and entities that should be invited to participate in the AC.

11. Assessment of the consequences of not doing an ACP.

12. Consideration of the expansion of other Sub-Area Plans and Response Plans beyond their current geographic area.

If the initial analysis concludes that there is little or no benefit from developing an ACP, or if there are insufficient resources to successfully conclude the project, then the RCP remains the operative plan and functions as the ACP.

If the conclusion of the initial analysis is that there would be a net benefit from developing an ACP, and that there are sufficient resources available (e.g., personnel, funding), the next step is to establish an AC. A lead agency or agencies should be designated to conduct outreach and provide information to potential AC members.

Before formally beginning the ACP process, it is wise to informally propose the concept to existing contacts in the response community to determine if they are supportive in general, and if they can commit time and resources. At this point, firm commitments are not required.

Area Committees

A. Initial AC recruitment

The first task is to identify the stakeholders and interest groups that should be involved in developing the ACP. Beginning with the list developed during the initial analysis, a summary of the analysis and an invitation to join the effort should be distributed by the lead agency. This first invitation list maybe supplemented with additional invitees as the effort gains visibility, with the objective to be as inclusive and comprehensive as possible. Potential members include:

❑ Existing RRT member agencies

❑ Other state/territorial/tribal agencies

  ● The lead state agency representative to the RRT should identify other state agencies with interest and expertise relevant to ACP development

❑ Regional and local agencies

  ● The LEPCs within the area should be the initial contact point for developing a list of potential participants

❑ NGOs

  ● This list should include such entities as industry associations and environmental organizations

❑ Private sector entities, including regulated facilities

  ● The potential list should include facilities required to have FRPs and others identified as high-risk in the initial analysis

Responses to the solicitation should be evaluated critically. Every invitee may not be able or willing to participate, so consideration must be given to identifying those that are most important to the success of the effort and finding incentives for them to participate. The number of AC members is variable and it is likely that several tiers of participation will emerge.

The AC recruitment phase may require several months of dialog with critical members that are reluctant to commit for various reasons (e.g., time constraints, limited resources). A decision must be made as to whether enough of the key members have committed to warrant proceeding with the initial AC meeting or whether additional preliminary work is needed.

B. Project management

Concurrent with the solicitation of participants, the lead agency should begin scoping the project because the costs and the timeframe will be important factors to potential AC members. It is unlikely that a definitive cost and schedule can be developed at this time, so an estimate should be developed based on previous experience. Other ACPs should be reviewed to find one that provides an appropriate model, and the AC for the selected model should be contacted to obtain information on realistic costs and schedules. This first estimate should include:

❑ Staffing requirements and costs

❑ Travel costs

❑ Contractor support requirements, sources, and costs

❑ Overall funding requirements and sources

❑ Time-line

C. Initial AC meeting

If the responses to the solicitation indicate sufficient interest by potential participating entities to proceed with the effort, an AC formation meeting should be scheduled and planned. The lead agency should prepare a briefing package for the meeting, including the initial analysis and the presumptive schedule and budget, along with related presentation materials. Briefing materials should be made available to invitees in advance.

The meeting date and location should be selected with care to afford maximum participation. Special consideration should be given to critical members.

Enlisting the support of state and local co-hosts for the meeting may help diffuse perceptions that the ACP is a federal 'top-down' project.

The essential elements of the initial meeting agenda include:

❑ Welcoming remarks by co-hosts

❑ Briefing by the lead agency on overall context, authorities, analysis, AC functions, schedule and budget

❑ Opportunity to review example plans from other areas

- ❑ Opportunity for clarifying questions and dialog
- ❑ Opportunity to opt for non-participation or information-only status
- ❑ Opportunity for potential members to describe their interests, authorities and resources
  - Discussion of obstacles (e.g., budgetary, bureaucratic)
  - Immediate follow-on meeting to plan for first formal meeting following the formation meeting

The first meeting of the AC should focus on obtaining a consensus on the following basics:

- ❑ AC membership and organization
  - Mail and email lists
  - Identification of other parties for further solicitation
  - Agreement on chair/co-chair agencies
- ❑ AC protocols
  - Meeting locations and scheduling, record-keeping, logistics
- ❑ Review of the initial analysis
  - Is the area covered by the ACP defined appropriately?
  - Are there errors or gaps in the initial analysis?
- ❑ General agreement on scope and schedule for ACP development
- ❑ Review of resources available for ACP development, as well as potential resource constraints
- ❑ Review of potential obstacles
- ❑ ACP format, focus, level of detail and distribution

Subsequent AC meetings will focus on the specific tasks of ACP development.

D. AC organization

There is no requirement that any specific organizational model be followed, and the AC may consider various options. The formality of the organization is also a function of its size; small ACs can be more informal and collegial, while large ACs may require a more clearly-defined structure.

The participating entities will likely fall into three groups:

- ❑ **Key players:** Those with an active interest and with sufficient resources to play an active continuing role in the AC.
- ❑ **Supporting players:** Entities with active, but limited interest or resources that may participate on specific issues.
- ❑ **Observers:** Entities that need to know about AC activities, but have no active role.

Organizationally, key players should be included in an executive committee or should act as chairs of potential subcommittees. Supporting players should be offered roles and positions tailored to their expertise and time/resource limitations. Observers may be kept informed through periodic reports and should be offered the opportunity to provide input and comment regarding elements of the ACP as they are developed.

E. AC operations

Participation in an AC is voluntary, and decision-making is primarily by consensus. EPA will chair the committee (with USCG as co-chair in areas where joint inland and coastal planning is conducted), but state and local participants may also be given co-chair roles. Each AC is headed by a Federal On-Scene Coordinator (FOSC); generally EPA personnel for the inland zone and USCG personnel for the coastal zone.

Member assignments are voluntary, and should be based on the interests, expertise and time/resources commitment required to execute the assigned tasks. Key AC operations are the following:

Review and approval processes: As tasks are completed, the AC must determine what levels of review are required, both within the AC itself and external to the AC. Draft ACP elements should be broadly distributed for review and comment, with reasonable deadlines (e.g., 30 days).When the AC determines that the ACP is complete, the ACP should be submitted to the involved EPA Regional Administrators (or their designees) for final approval. (See Appendix A, April 24, 1992 *Federal Register* Notice.)

Documentation, record-keeping and administrative support: An important early AC decision is to determine the appropriate level of record-keeping and documentation, and the amount of administrative support required to maintain this level. Administrative support may be provided by staff from participating agencies or from support contractors, and includes the maintenance of files, distribution lists, web sites and other tasks.

Reference materials library and distribution: As the ACP process evolves, the set of reference materials supporting the ACP will grow. Maintaining this set and ensuring that all participants have access to it is a separate administrative support task.

Membership management: The most important part of this task is the maintenance of the AC membership and distribution lists, ensuring that mail addresses, email addresses and phone numbers are kept accurate and current. For those members of the AC with specific AC responsibilities, back-up or alternate members should also be identified to ensure continuity is maintained when a member is unavailable.

F. AC activities and responsibilities

The lead Agency designated FOSC is responsible for developing and managing the ACP through the AC. These tasks include:

❑ **ACP development:** The AC's primary task is to produce a final ACP that meets the needs defined by the AC members and achieves the standards required.

- ❑ **ACP publishing and distribution:** Once the final ACP is developed, the AC should decide on the modes of publication (e.g., hard copy, electronic, internet) and the scope of distribution.

- ❑ **ACP evaluation:** Once the ACP is distributed, the AC should establish a mechanism for evaluating its effectiveness. The ACP itself should contain contact addresses for receiving feedback and the AC should periodically consider comments received for possible future amendments. In addition, the AC may participate in after-action analyses of significant incidents and exercises.

- ❑ **ACP updates and modifications:** Technological advances, jurisdictional and organizational changes, infrastructure changes and other factors may require the ACP to be modified and updated. The AC should consider establishing an appropriate update cycle. In addition, a means for providing interim updates should be established for significant events that cannot be deferred to the update cycle. This makes ACP version control and date-stamping an essential part of ACP management and enables all users to work with the most current information.

- ❑ **Inreach and outreach activities:** At a minimum, the AC may be requested to provide reports on progress to the RRT. Beyond that, the AC should consider how information should be disseminated among the area community. The AC may serve as a clearing-house for planning and response-related news. AC members should inform their own agencies about the ACP and how to access and use it, so that all responders are equally prepared when an incident occurs. The AC should also be prepared to respond to requests for information from outside entities and organizations; this may require the preparation of fact sheets and briefing materials that provide a general overview of the ACP.

- ❑ **ACP Environmental Benefits Analysis:** Net Environmental Benefits Analysis (NEBA) is a methodology for identifying and comparing environmental benefits of alternative management options in the removal of discharged oil and oil products. Net environmental benefits are the gains in environmental services or other ecological properties attained by the removal of the oil or ecological restoration minus the environmental injuries caused by those actions. A NEBA for oiled sites typically involves the comparison of the following management alternatives:

  - Leaving contamination in place for natural attenuation;
  - Removing the contaminants through traditional removal techniques; and
  - Remediating contamination with alternative removal techniques.

NEBA involves valuing ecological services or other properties, assessing adverse impacts, and evaluating removal actions. NEBA is a risk-benefit analysis applied to environmental management options. To do this, a balance of resource managers and emergency responders must participate together in forming opinion, guiding discussion and educating each other in processes of importance and concern.

NEBA has the potential to assist resources managers avoid the possibility that the selected removal alternative will provide no net environmental benefit over natural attenuation of contaminants and ecological recovery. An alternative removal option may provide no net environmental benefit because:

- The removal action is ineffective or inappropriate (the action does not substantially change the risk); or

- The removal alternative causes environmental injuries greater than the damage associated with the contamination because:

  - The need for remediation has been driven by human health risk, not ecological risk;

  - The ecological injury from contamination has been overestimated;

  - Injuries associated with removal were not properly addressed; or

  - The need for remediation is driven by human considerations not related to health or ecologic concerns.

NEBA has the potential to help resource managers plan a removal that provides a positive net environmental benefit over the hypothetical state that would prevail in the absence of contamination. NEBA may be appropriate if any of the removal alternatives potentially have significant negative ecological effects or minimal ecological benefits. Finally, NEBA may be used when the multiple alternatives are beneficial, but the one with the greatest net benefits is not apparent without formal analysis.

See Appendix D for examples of sensitive areas inventories.

❑ **ACP-based drills and exercises:** The AC itself may sponsor drills and exercises to test the viability of ACP elements and it should encourage its members to use the ACP when conducting such activities in internal agency venues and in inter-agency exercises. The AC should ensure that exercise evaluations relating to the ACP are included in the ACP improvement process.

❑ **ACP-related training:** The AC may develop and sponsor training activities to improve the ability of responders to access and utilize the ACP. These may be specific to the ACP or may include more general topics, such as NIMS-ICS courses, Health and Safety courses or spill response courses. For electronic and web-based ACPs, the AC should consider the need for training in the use of software that may be required for access and utilization.

For examples of AC organizations, documents, agendas and processes, see Appendix C.

Section 4

Scope and Content of the ACP

❑ **ACP coverage:** The area covered by the ACP may be defined by geographic features, jurisdictional boundaries, or both, at the discretion of the AC. Within the ACP boundaries, sub-areas may be defined where there are unique circumstances that require tailored response strategies.

❑ **Areas of special economic and environmental importance:** The ACP must include an inventory of features within the area that require awareness by responders when developing response strategies.

- *Critical infrastructure:* Utilities (such as drinking water intakes, water and wastewater treatment plants, and major electrical power plants and transmission lines), transportation corridors and facilities, and other infrastructure elements may require specific protection measures, special notification or access protocols or have other unique attributes that may affect a response. The ACP should identify these features and also provide guidance on how they should be considered in response strategies.

- *Environmentally sensitive areas:* The ACP should identify areas within its bounds that may require tailored protection or response strategies due to unique environmental attributes. These may be recreational or commercially-significant areas, endangered species habitats, drinking water supplies or other areas defined by the AC. In each case, the ACP should provide guidance on how responders should incorporate the needs of these areas into response strategies.

- *Culturally sensitive areas:* The ACP should identify historical landmarks, archeological sites, tribal lands and other features that may require special protective measures or interaction with trustees or tribal authorities.

- *High-risk locations:* The ACP should identify fixed facilities and transportation infrastructure locations that present a high risk of release of oil or hazardous substances. Once these are identified, the ACP should then address location-specific response strategies and preparedness, such as the pre-staging of response equipment. To the extent that these locations may be subject to regulatory requirements, such as facilities required to have an FRP, the ACP should reference or provide a link to the FRP or other required plan. The ACP should also reference and, whenever possible, link to plans and other information developed by LEPCs within its bounds.

- *Natural disaster impact areas*: The ACP should incorporate information relating to locations that may be susceptible to natural disaster impact (e.g., flooding, earthquakes), and provide references or links to related disaster response plans at the local, state and federal levels.

❑ **Identifying and integrating with other plans:** The ACP should identify and define its relationship to other contingency plans that are within, adjacent to, or overlapping the ACP defined area. These plans should be reviewed to ensure the ACP is not inconsistent with them, and the owners of these plans should be informed of the ACP's status and receive copies of the final ACP. If there is a reasonable prospect of an incident occurring that impacts both the ACP area and an area covered by an adjacent plan, the AC should establish notification and coordination protocols with the adjacent entities. Examples of other plans to consider include:

- *External plans*: Adjacent RCPs and ACPs and International border plans

- *Internal plans*: State and local plans, private sector plans (FRPs and Risk Management Plans (RMPs))

❑ **Overall ACP formats:** The AC should review example formats for the ACP to determine the most appropriate fit for the needs of area responders. The primary purpose of the ACP is to serve as a response tool. The primary customers of the AC process are the area responders, so the ACP must be portable, easy to navigate, and accurate.

❑ **ACP maintenance:** Once the ACP is issued in final form, the AC should implement a management and maintenance process to keep the ACP current and to incorporate improvements. A regular update cycle should be considered to provide for changes that are not time-critical, but interim amendments may also be required to reflect significant changes within the defined area. Version control should be established and an interim update process is critical. Certain portions of the plan, such as contact lists, may change frequently and should be maintained separately from the plan itself.

❑ **Downloadable and Internet-access ACPs:** Consideration should be given to distribution of the ACP in electronic form, to usability on PDAs and PCs, and to providing access to the ACP via internet.

Essential Plan Elements

A. Maps

Maps are central to ACP development and utilization. The variety of mapping formats, platforms and applications is constantly increasing and evolving. Mapping tools should be evaluated in terms of accuracy, accessibility, usability for responders in the field and ease of maintenance and updating. Appendix D provides links to mapping tools that have proven useful in ACP development.

B. Contacts and notification

Contact and notification lists must be maintained for a variety of purposes. These lists may include:

❑ Lists relating to the AC and the ACP itself, the first of which includes individuals and entities engaged in ACP development and maintenance. Other lists may cover those that receive ACP copies for information only.

❑ Lists related to response operations should cover both immediate notifications when an incident occurs and contacts during response operations when assistance is required from an entity listed in the plan with knowledge, authority, expertise or resources required by the Incident Commander (IC)/Unified Command (UC). In general, these lists should be maintained separately from the plan itself, since they may contain information that should not be widely disseminated. Lists of this type are not intended to supplant existing notification protocols, but reinforce and supplement them by adding information specific to the area covered by the ACP. Response operations lists should include 24/7 contact information for all essential response entities.

❑ All lists should include mail, email, land line, and cell phone contact information, as well as back up numbers if the primary contact is unavailable and general agency office numbers.

❑ List management should be through a central administrative support control point.

C. Resources

The resources section of the ACP is perhaps the most difficult to develop and manage. This is primarily due to the sheer magnitude of developing an inventory of the personnel, equipment and capabilities of all response

entities in the defined area, and to the difficulty of keeping the inventory current. It is advisable that the AC initially limit its level of comprehensiveness, detail and focus on the general capabilities of the response entities and on those resources that are unique and may be difficult to acquire. If the Incident Commander (IC)/Unified Command (UC) requires specific information from a response entity, contact points can be provided to obtain the most current information directly from the provider.

Resource information is organized at the discretion of the AC. The information may be organized by resource category, agency, type of incident or some combination of these or other categories. Regardless of the organization, there are minimum requirements that should be incorporated into the inventory. After determining the organizational concept, the AC should identify the initial resource requirements and establish a spreadsheet format for agencies to enter their resource information.

❑ **General capabilities:** Each entity identified as potentially having a response or response support role should describe its authorities, areas of jurisdiction, areas of expertise, types of available personnel and equipment and general response capabilities, including access to funds.

❑ **Personnel:** The inventory may include numbers of available personnel, field-deployment qualifications (including OSHA qualifications), Incident Command System (ICS) qualifications, areas of technical and scientific expertise, mobilization response times, non-deployable support personnel, secondary resources (available through contracts or mutual aid/ Emergency Management Assistance Compact (EMAC) agreements), and any other criteria that the AC identifies as necessary.

❑ **Equipment:** Subcategories may include assessment, soil/water/air sampling, field categorization, ambient monitoring, aerial survey/remote sensing, transportation, field logistics, transportation, heavy equipment, booms, pumps, skimmers, PPE, mobile command posts (MCPs), communications, data management.

❑ **Laboratories:** Identification of which entities have access to analytical capability, general descriptions of capability and access procedures and contact points.

❑ **Volunteer Resources:** Management of volunteer resources presents unique issues regarding training, safety, liability and integration with the response organization. The ACP should provide links to policy documents relating to volunteer management and to local organizations with volunteer management expertise. The NRT has developed guidance to address these issues (See Appendix I).

In applying this guidance to the ACP, the AC should insure that the ACP includes an inventory of potential volunteer organizations, with brief descriptions of their interests, capabilities and contact information. To develop this list, the AC should task a work-group to conduct outreach to volunteer organizations to inform them of ACP activities and the parameters for response participation, including training, safety and liability management requirements and to identify potential obstacles to successful integration of volunteers into the response organization. The AC should also consider the unique issues involving the use of volunteers during a response. See Use of Volunteers, National Oil and Hazardous Substances Pollution Contingency Plan (NCP), 40 Code of Federal Regulations (CFR) § 300.5, located in Appendix I of this Handbook.

If the AC determines that volunteer management may be a significant factor in responses, then additional actions may be needed, such as inviting volunteer organizations to participate as members of the AC, developing advance and/or just-in-time training programs in NIMS/ICS, safety (e.g., HAZWOPER) and technical response subjects (e.g., wildlife rehabilitation) and inviting volunteers to participate in ACP-related exercises.

❑ **Contact information:** Contacts for each type of resource, including level of approval needed for commitment.

Information from each inquiry should be entered into a searchable database so that potential resources can be identified quickly.

D. Sensitive areas

The AC should establish a committee to identify features and sub-areas that are sensitive for environmental, cultural or economic reasons. This committee should include entities with expertise in the application of requirements established by the Endangered Species Act (ESA), the Historical Preservation Act and other statutes, regulations and agreements concerning sensitive areas. The common theme for identifying a sensitive area is that it has attributes that must be considered by responders in developing response strategies and tactics. For each feature or area identified, the exact location or boundaries should be mapped when possible, and a brief summary of considerations should be documented. For certain sensitive areas, such as ESA or archaeological sites, exact locations may not be identified, but should be referenced as present in the general area. This summary should include:

❑ Specific attributes (e.g., drinking water supply intake, endangered species habitat)

❑ Recommendations on protective measures that may be employed

❑ Description of any proscribed tactics

❑ Contact information for operators, trustees and others with an interest in the sensitive area

❑ Other information relevant to the area, such as special access protocols, hazards to responders or seasonal variations to be considered in developing response strategies and tactics

This information is organized at the discretion of the AC and is dependent on the available data. Consideration should be given to assigning priority categories to sensitive areas, based on their significance, or organizing them by type. Areas with especially difficult or complex issues should be considered for development of specific Sub-area Plans or Geographic Response Plans.

The AC should consider whether certain types of information should preliminarily be designated for restricted use only. Each AC must communicate with the "owner" of the information and determine if their information falls in this category and if so, how the information will be safeguarded but available during an emergency response. All records featuring such information may ultimately be subject to public disclosure, however, in response to a FOIA request.

The ACP should ensure that the appropriate state, Federal, and tribal trustees for natural resources are promptly notified of discharges and response activities are coordinated with the with the affected natural resource trustees. Additional information on notification and coordination with natural resource trustees is available at *http://www.epa.gov/superfund/programs/nrd/trustees.htm*.

Tools for identifying sensitive areas: See Appendix D: Tools.

Methods for organizing sensitive area data: See Appendix D: Tools.

Methods for display and accessing data: See Appendix D: Tools.

E. Hazard analysis

The AC should establish a sub-committee to identify potential sources of releases within the defined area. These sources may include fixed facilities or transportation routes with high volumes of oil or hazardous materials in transit. Consideration should also be given to potential sources outside the defined area of the ACP, which may impact the area in the event of a release The first task of the committee is to develop working criteria to establish a cut-off point, below which potential sources will not be addressed by the ACP. These need not be rigid; for example, if potential sources A and B are otherwise identical, but A is within a defined sensitive area, the ACP may address A and leave B below the threshold.

For each potential source identified the ACP should document the following:

❑   Source location (to be mapped)

- Operator, with contact and access information
- Types and quantities of materials that may be released
- Special considerations for responders, including hazards
- Response capabilities of the operator

Tools for identifying potential sources:

- FRPs: EPA FRP Coordinators list posted at *http://www.epaosc.org/site/site_profile.aspx?site_id=3857*.

- Pipelines: The U.S. Department of Transportation (DOT) Pipeline and Hazardous Materials Safety Administration (PHMSA) web site (*http://www.phmsa.dot.gov/*) includes a range of pipeline safety resources, including a national pipeline mapping system.

- Railroads: The DOT Federal Railroad Administration (FRA) web site (*http://www.fra.dot.gov/*) includes passenger and freight railroad safety and environmental information. The FRA's GIS web site provides a web-based mapping application that permits users to map, view and zoom to all rail grade crossings in the United States. Accident information for each grade crossing is available.

- Highways: The DOT Federal Motor Carrier Safety Administration (FMCSA) maintains a Hazardous Materials Routing Web Site that lists designated, preferred and restricted routes (*http://www.fmcsa.dot.gov/safety-security/hazmat/hm-theme.htm*).

- Hazmat facilities: Facilities covered by EPCRA requirements must submit an Emergency and Hazardous Chemical Inventory Form to the LEPC, the State Emergency Response Commission (SERC), and the local fire department annually. Facilities provide either a Tier I or Tier II form. Most states require the Tier II form. Some states have specific requirements in addition to the Federal Tier II requirements. The EPA web site includes a list of links to state Tier II reporting sites: *http://www.epa.gov/osweroe1/content/epcra/tier2.htm*. Tier II data for most states are also maintained on the E-Plan Emergency Response Information System: *https://erplan.net/eplan/login.htm*.

EPA's Toxics Release Inventory (TRI) is a database containing data on releases of over 600 toxic chemicals from thousands of U.S. facilities and information about how facilities manage those chemicals through recycling, energy recovery, and treatment. One of TRI's primary purposes is to inform communities about toxic chemical releases to the environment. TRI data are available at *http://www.epa.gov/tri/*.

The Facility Registry System (FRS) is a centrally managed database developed by EPA's Office of Environmental Information (OEI) that identifies facilities, sites or places subject to environmental regulations or of environmental interest. FRS creates high-quality, accurate, and authoritative facility identification records through rigorous verification and management procedures that incorporate information from program national systems, state master facility records, data collected from EPA's Central Data Exchange registrations and data management personnel. The FRS provides Internet access to a single integrated source of comprehensive (air, water, and waste) environmental information about facilities, sites or places. FRS data are available for query at *http://www.epa.gov/enviro/html/fii/index.html*.

Hazardous waste generators, transporters, treaters, storers and disposers of hazardous waste are required to provide information on their activities to state environmental agencies. These agencies then provide the information to EPA offices through the Resource Conservation and Recovery Act Information (RCRAInfo) System (*http://www.epa.gov/enviro/facts/rcrainfo/search.html*). Information on cleaning up after accidents or other activities that result in a release of hazardous materials to the water, air or land must also be reported through RCRAInfo.

Superfund is a program administered by the EPA to locate, investigate, and clean up uncontrolled hazardous waste sites throughout the U.S. CERCLIS Search is available to retrieve Superfund data from the Comprehensive Environmental Response, Compensation, and Liability Information System (CERCLIS) database in Envirofacts (*http://www.epa.gov/enviro/facts/cerclis/search.html*).

☐ **LEPC plans:** Information on LEPCs can be found at *http://www.epa.gov/oem/content/epcra/epcra_plan.htm#LEPC*.

☐ **RMPs:** The Right-To-Know Network maintains a Risk Management Plan (RMP) Database on its web site (*http://www.rtknet.org/db/rmp*). RMP information may also be accessed at Federal Reading Rooms: *http://www.epa.gov/oem/content/rmp/readingroom.htm*.

Natural Disaster-sensitive areas and facilities: The AC should also consider which significant facilities may be vulnerable to impact by natural disasters, such as floods or earthquakes.

Methods of organizing potential source information: See Appendix D: Tools

Methods for displaying and accessing data: See Appendix D: Tools

F. Response strategies and worst-case discharges

After the AC has developed the initial inventory of sensitive areas and potential sources, it can begin to consider the general response strategies with special consideration given to potential worst case discharges.

☐ **Assessment strategies**

The AC will have to identify methods to assess the extent and impact of a release and identify the tools available to predict the behavior of released material. Remote sensing, modeling and sampling strategies should be developed as needed.

☐ **Protection strategies**

The AC should determine the most effective methods of preventing impact on sensitive areas.

❑ Response strategies

The ACP should identify the various response strategies that have proven to be effective in controlling and mitigating the impact of a release. Consideration should be given to the most likely release scenarios and the worst-case discharges.

❑ Oil-spill-specific strategies and plans (e.g., the NRT Subsea Dispersant Guidance) including counter-measures

Oil spill counter-measures include dispersants[1], in-situ burning (including accelerants), bio-remediation, surface washing agents, solidifiers and other methods for reducing the impact of oil to the environment. While many of the countermeasure stipulations are included in the RCPs, the ACP must also consider counter-measure use in the context of the defined ACP area. These issues include:

- Areas where specific counter-measures may be proscribed
- Pre-approval of specific counter-measures in certain areas
- Protocols for monitoring use and effectiveness
- Assessment of potential impacts from counter-measure use in adjacent planning areas (e.g., Coastal Zone areas)

❑ Facility-specific strategies and plans

Facilities with the potential for large-scale releases (such as pipelines and large storage and manufacturing facilities and railroads) should be considered for focused strategy development. If facilities are covered by FRPs, the plans will provide a base for the responding agencies to develop strategies for most-likely and worst-case releases from these facilities.

G. Response management: roles and responsibilities

❑ NIMS compliance policy: The ACP should include a brief section that commits the AC to NIMS compliance and references the Incident Management Handbooks and Field Operating Guides that are used by participating agencies.

❑ Unified Command: One of the most important functions of the ACP is to address potential jurisdictional conflicts and to provide solutions to these in advance of a response. This section should identify the agencies that meet the criteria for participating in a UC, including appropriate jurisdictional authority, ability to commit resources to the response, and personnel that are trained and qualified to serve as Incident Commanders. Consideration should also be given to the role of responsible parties in the UC. In areas where there are multiple overlapping jurisdictions, this task may need to be broken down into scenario-based organizations.

---

1 Dispersants or other oil emulsifiers are not utilized in freshwater and other inland environments because of the limited dilution available in fresh waters, the use of freshwaters as a water supply, the limited toxicology information available for dispersants in fresh water, and the limited information available as to fresh water effectiveness of dispersants. As of 2012, there are no dispersants that are effective in freshwater environments (dispersants require salt as a catalyst).

❑ **Response Organizations:** This section should provide guidance on NIMS-compliant response organizations, identifying those entities with expertise relevant to specific positions and providing models of organizational structures. The approach to this should be inclusive, by defining appropriate roles for each AC participant. Particular attention should be paid to the placement of resource trustees, technical experts and others that may be outside the normal response community.

❑ **Personnel training and qualification requirements/recommendations:** This section should address recommended levels of NIMS-ICS training for responders.

❑ **Model Incident Action Plans:** It may be appropriate for the ACP to include example IAPs for specific scenarios (e.g., worst-case discharges).

❑ **Mutual aid agreements:** These exist at the federal, state and local levels. Federal agreements (e.g., EPA-Coast Guard) and state agreements (e.g., EMAC) need not be replicated in the ACP unless there are area-specific considerations which need to be explained. Local agreements, particularly when they involve entities outside the bounds of the ACP, should be referenced briefly.

❑ **Public Information/JIC:** The ACP should provide guidance to participants on the coordination of public messages during a response, including reinforcing the role of the UC's PIO and defining the relationship of the PIO to individual agencies' public information operations.

❑ **Response to substantial threats to public health or welfare; spills of national significance and worst case discharges:** As described in 40 CFR 300.322 through 300.324, if the investigation by the OSC shows that the discharge poses or may present substantial threat to public health or welfare of the United States, the OSC shall direct all federal, state or private actions to remove the discharge or to mitigate or prevent the threat of such discharge, as appropriate.

The ACP, when used in conjunction with other provisions of the NCP, shall be adequate to remove worst case discharges as described above.

Advanced Area Planning

A. NOAA Environmental Response Management Application (ERMA)

ERMA® is a web-based Geographic Information System (GIS) tool designed to assist both emergency responders and environmental resource managers who deal with incidents that may adversely impact the environment. The application can assist in response planning and is accessible to both the command post and to assets in the field during an actual response incident, such as an oil spill or hurricane. The data within ERMA also assists in resource management decisions regarding hazardous waste site evaluations and restoration planning.

ERMA supports environmental preparedness, response, and recovery efforts by:

❑ Providing integrated and timely information to improve decision-making.

❑ Integrating and synthesizing various types of information on a single map interface.

❑ Providing fast visualization of current information.

❑ Improving communication and coordination among responders and stakeholders.

Access information for ERMA is available at *http://response.restoration. noaa.gov/erma* and is also included in Appendix D.

B. Computer-Aided Management of Emergency Operations (CAMEO)

CAMEO is a system of software applications used widely to plan for and respond to chemical emergencies. It is one of the tools developed by EPA's Office of Emergency Management (OEM) and the National Oceanic and Atmospheric Administration (NOAA) Office of Response and Restoration to assist front-line chemical emergency planners and responders. They can use CAMEO to access, store, and evaluate information critical for developing emergency plans. In addition, CAMEO supports regulatory compliance by helping users meet the chemical inventory reporting requirements of the Emergency Planning and Community Right-to-Know Act (EPCRA, also known as SARA Title III). CAMEO also can be used with a separate software application called LandView to display EPA environmental data and demographic/economic information to support analysis of environmental justice issues.

The CAMEO system integrates a chemical database and a method to manage the data, an air dispersion model, and a mapping capability. All modules work interactively to share and display critical information in a timely fashion. The CAMEO system is available in Macintosh and Windows formats.

CAMEO was initially developed because NOAA recognized the need to assist first responders with easily accessible and accurate response information. Since 1988, EPA and NOAA have collaborated to augment CAMEO to assist both emergency responders and planners. CAMEO has been enhanced to provide emergency planners with a tool to enter local information and develop incident scenarios to better prepare for chemical emergencies. The Bureau of Census and the U.S. Coast Guard have worked with EPA and NOAA to continue to enhance the system.

The software is available for download from EPA's CAMEO web site:
http://www.epa.gov/osweroe1/content/cameo/index.htm

C. LandView® 6

The LandView database system allows users to retrieve census demographic and housing data, EPA Envirofacts data and U.S. Geological Survey (USGS) Geographic Names Information System (GNIS) information. The GNIS contains over 1.2 million records which show the official federally recognized geographic names for all known places, features, and areas in the United States that are identified by a proper name.

The LandView database software:

❑ Uses the Population Estimator function to calculate census demographic and housing characteristics for user defined radii.

❑ Creates simple thematic maps of census data.

❑ Allows users to browse and query the census, EPA or USGS databases and show the query results on the map.

❑ Provides the capability to locate a street address or intersection on a map based on TIGER/ Line® road features and address ranges.

The MARPLOT mapping software:

- ❑ Creates large scale maps showing Census legal and statistical entities, EPA Envirofacts sites, and USGS GNIS features. (A large scale map shows a small area with a large amount of detail.)

- ❑ Allows users to customize the maps by varying the scale and controlling which map layers are shown.

- ❑ Provides a search capability for map objects based on radius or map layer.

- ❑ Includes tools that allow users to add information to the maps.

- ❑ Can automatically retrieve LandView database information for user selected map objects.

Additional information is available at: *http://www.census.gov/geo/landview/*

D. RMP*Comp

RMP*Comp is a free program that calculates vulnerable zone distances based on the Risk Management Program (RMP) Guidance for Offsite Consequence Analysis (both worst case scenarios and alternative scenarios). The RMP*Comp program guides users through the process of making an analysis.

The software is available for download from EPA's RMP*Comp web site: *http://www.epa.gov/osweroe1/content/rmp/rmp_comp.htm*

This page intentionally left blank

Appendix A:
Statutory and Regulatory Authorities

CERCLA and EPCRA

Under CERCLA, EPA established both an emergency response program designed to stabilize or cleanup releases of hazardous substances that pose a threat to human health or the environment, and a remedial response program to take actions consistent with a permanent remedy (instead of or in addition to removal actions) in the event of a release or threatened release of hazardous substances posing a threat to human health or the environment. CERCLA also enabled the revision of the NCP. The NCP provides the guidelines and procedures to respond to releases and threatened releases of hazardous substances, pollutants, or contaminants. The Emergency Planning and Community Right-to-Know Act (EPCRA) amendments to CERCLA included provisions to strengthen emergency response planning at the state and local levels by requiring local governments to prepare chemical emergency response plans (40 CFR Part 355) and to make information more readily available to the public on hazardous chemicals that are stored at facilities in their communities (40 CFR Part 370).

Clean Water Act

Under 33 U.S.C. 1321 (j)(4) of the CWA, the President is authorized to establish Area Committees comprised of qualified personnel from federal, state, and local agencies. These committees are to prepare ACPs that detail methods and procedures for responding to a worst-case discharge, including the division of responsibilities among various authorities in a response. Each Area Committee is required to submit this plan to the President for review and approval. The authorities assigned to the President under 33 U.S.C. 1321(j)(4) for the inland zone have been delegated by Executive Order 12777 to the EPA Administrator, who has in turn re-delegated these authorities to EPA Regional Administrators. Regional Administrators may further re-delegate the authorities to the Division Director level.

Each Area Committee, under the direction of the Federal On-Scene Coordinator (FOSC) for its area, has the following responsibilities:

- Prepare an ACP for its area;

- Work with state and local officials to enhance the contingency planning of those officials and to assure pre-planning of joint response efforts, including appropriate procedures for mechanical recovery, application of countermeasures, shoreline cleanup, protection of sensitive environmental areas, and protection, rescue, and rehabilitation of fisheries and wildlife;

- Work with state and local officials to expedite decisions for the use of dispersants and other mitigating substances and devices; and

- Update the ACP periodically.

The Oil Pollution Act of 1990 (OPA 90)

OPA 90 establishes mechanisms for the federal government to prevent and respond to oil spills. OPA 90 extensively amended the CWA to provide enhanced capabilities for oil spill response and natural resource damage assessment.

Title IV, Section 4202, National Planning and Response System, amended subsection 311(j) of the CWA with respect to the National Planning and Response System. It defines Area Committee and ACP requirements and deadlines for agencies. Pursuant to OPA 90 section 4202(b)(1)(A), the President designates areas for which ACPs are established. As stated above, the President delegated to EPA the responsibility for designating the areas and appointing the committees for the "inland zone". Under the CWA, ACPs are developed by Area Committees under the direction of the FOSC for their area. OPA 90 Section 4202(b)(1)(A) also requires that in designating areas, the President will ensure that all navigable waters, adjoining shorelines, and waters of the exclusive economic zone are subject to an ACP.

Under the National Oil and Hazardous Substances Contingency Plan (NCP) response and planning framework, the territory of the United States is covered by thirteen Regional Response Teams (RRTs) and Regional Contingency Plans (RCPs). The zones of the thirteen RRTs follow the ten standard federal regions, except for the following three subregional areas that have their own RRT: (1) Puerto Rico and the U.S. Virgin Islands; (2) Alaska; and (3) Hawaii, Guam, Northern Mariana Islands, Pacific Island Governments, and American Samoa (See Figure 1, next page). The inland areas of the thirteen RRTs serve as the designated areas for the inland zone. The USCG designates areas for the

Coastal Zone. These coastal zone areas are based on the 48 USCG Captains of the Port (COTP) areas. The areas covered by COTPs are smaller than the RRT areas and include major river systems associated with the ports.

Unless otherwise designated, the RRTs serve as the Area Committees for the Inland Zone. RRTs are composed of representatives from federal, state, and Tribal governments.

See also the April 24, 1992 *Federal Register* Notice (57 FR 15198): Designation of Areas and Area Committees Under the Oil Pollution Act of 1990 (Document posted at *http://www.epaosc.org/site/site_profile. aspx?site_id=3857*)

The National Oil and Hazardous Substances Pollution Contingency Plan (NCP)

The National Oil and Hazardous Substances Pollution Contingency Plan (NCP) provides for the coordinated and integrated response by the federal government, as well as state and local governments, to prevent, minimize, or mitigate a threat to public health or welfare posed by discharges of oil and releases of hazardous substances, pollutants, and contaminants. The NCP is authorized by CERCLA and the CWA as amended by OPA 90.

Section 300.210 of the NCP provides for three levels of contingency plans under the NRS, including: The NCP, Regional Contingency Plans (RCPs), and ACPs. These plans are available for inspection at EPA Regional offices or USCG district offices.

Under the direction of a FOSC and subject to approval by the lead agency, each Area Committee, in consultation with the appropriate RRTs, USCG District Response

Figure 1
13 Regional Response Team Areas

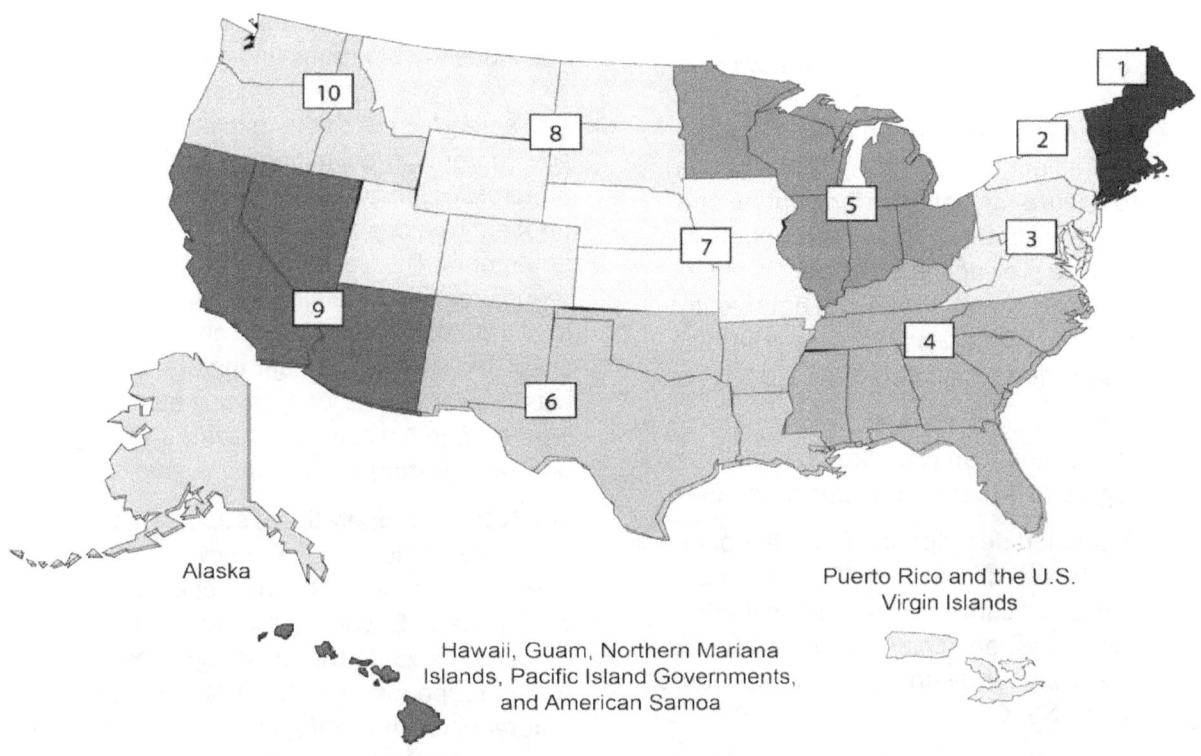

Alaska

Puerto Rico and the U.S. Virgin Islands

Hawaii, Guam, Northern Mariana Islands, Pacific Island Governments, and American Samoa

Groups (DRGs), the USCG National Strike Force Coordination Center (NSFCC), Scientific Support Coordinators (SSCs), Local Emergency Planning Committees (LEPCs), and State Emergency Response Commissions (SERCs), will develop an ACP for its designated area. This plan, when implemented in conjunction with other provisions of the NCP, will be adequate to remove a worst-case discharge of the NCP, and to mitigate or prevent a substantial threat of such a discharge, from a vessel, offshore facility, or onshore facility operating in or near the area.

In developing the ACP, the FOSC will coordinate with affected SERCs and LEPCs. The ACP will provide for a well-coordinated response that is integrated and compatible, to the greatest extent possible, with all appropriate response plans of state, local, and non-federal entities, and especially with Title III local emergency response plans.

The ACPs are required to include the following elements:

❑ A description of the area covered by the plan, including the areas of special economic or environmental importance that might be damaged by a discharge;

❑ A detailed description of the responsibilities of an owner or operator and of federal, state, and local agencies in removing a discharge, and in mitigating or preventing a substantial threat of a discharge;

- ☐ A list of equipment (including firefighting equipment), dispersants, or other mitigating substances and devices, and personnel available to an owner or operator and federal, state, and local agencies, to ensure an effective and immediate removal of a discharge, and to ensure mitigation or prevention of a substantial threat of a discharge (this may be provided in an appendix or by reference to other relevant emergency plans (e.g., state or LEPC plans), which may include such equipment lists);

- ☐ A description of procedures to be followed for obtaining an expedited decision regarding the use of dispersants; and

- ☐ A detailed description of how the plan is integrated into other ACPs and tank vessel, offshore facility, and onshore facility response plans approved by the President, and into operating procedures of the NSFCC.

Area Committees are required to incorporate into each ACP a detailed annex containing a Fish and Wildlife and Sensitive Environments Plan that is consistent with the RCP and NCP. The annex will be prepared in consultation with the U.S. Fish and Wildlife Service, the National Oceanic and Atmospheric Administration (NOAA), and other interested natural resource management agencies and parties. The annex will provide the necessary information and procedures to immediately and effectively respond to discharges that may adversely affect fish and wildlife and their habitat and sensitive environments, including provisions for a response to a worst-case discharge.

The Stafford Act

The Robert T. Stafford Disaster Relief and Emergency Assistance Act (Stafford Act) describes the programs and processes by which the federal government provides disaster and emergency assistance to state and local governments, tribal nations, eligible private nonprofit organizations, and individuals affected by a declared major disaster or emergency. The law establishes the process for requesting and obtaining a Presidential disaster declaration, defines the type and scope of assistance available under the Stafford Act, and sets the conditions for obtaining assistance. The Stafford Act covers all hazards, including natural disasters and terrorist events.

The NCP is an operational supplement to the National Response Framework (NRF). The NRF was issued by the Department of Homeland Security (DHS) and is an overarching guide that describes how the nation responds to all types of domestic emergencies, including natural disasters and terrorist incidents. It describes the roles of federal, state, local, and tribal governments, as well as non-governmental organizations and the private sector. Under the NRF, DHS coordinates the federal response to incidents requiring significant Federal coordination, which includes incidents for which the President issues a disaster declaration under the Stafford Act. FEMA may utilize Stafford Act funds to reimburse EPA for specific emergency response activities related to actual or potential hazardous materials (hazardous substances, pollutants, contaminants, and oil) incidents through the NRF under Emergency Support Function (ESF #10) – Oil and Hazardous Materials Response, when there is an Emergency or Major Disaster Declaration.

Response to oil and hazardous materials incidents is generally carried out in accordance with the NCP. NCP structures and response mechanisms remain in place when ESF #10 is activated, but coordinate with NRF mechanisms. During Stafford Act responses, some procedures in the NCP may be streamlined or may not apply.

ESF #10 may be activated by DHS for incidents requiring a more robust coordinated Federal response, such as:

❏ A major disaster or emergency under the Stafford Act;

❏ A federal-to-federal support request (e.g., a federal agency, such as the Department of Health and Human Services (HHS) or U.S. Department of Agriculture (USDA), requests support from ESF #10 and provides funding for the response through the mechanisms described in the Financial Management Support Annex); or

❏ An actual or potential oil discharge or hazardous materials release to which EPA and/or USCG respond under CERCLA and/or CWA authorities and funding, for which DHS determines it should lead the federal response.

As described in the NRF core document, some federal responses do not require coordination by DHS and are undertaken by other federal departments and agencies consistent with their authorities. Federal responses to oil and hazardous materials incidents under the authorities of CERCLA and the CWA that do not warrant DHS coordination are conducted under the NCP. EPA or USCG may also request DHS to activate other NRF elements for such incidents, if needed, while still retaining overall leadership for the federal response.

Homeland Security Presidential Directives (HSPD)/Presidential Policy Directives (PPD)

The Presidential Policy Directive on National Preparedness is effective as of March 30, 2011, and replaces Homeland Security Directive (HSPD) 8 (National Preparedness) and HSPD-8 Annex I (National Planning). Plans developed under HSPD-8 and Annex I remain in effect until rescinded or otherwise replaced.

EPA has also been directed to operate under the following Executive Branch directives:

Management of Domestic Incidents – HSPD-5

HSPD-5 was issued to improve management of domestic incidents by establishing a single, comprehensive national incident management system. The Homeland Security Act of 2002 created the Department of Homeland Security (DHS) and assigned the Secretary of Homeland Security responsibility for coordinating federal emergency operations within the United States. Federal emergency operations include preparing for, responding to, and recovering from terrorist attacks, major disasters, and other emergencies. DHS has the authority to coordinate federal resources when any one of several conditions occurs: 1. a federal department or agency requests their assistance, 2. the resources of state and local authorities are overwhelmed and they request federal assistance, 3. more than one federal department or agency is substantially involved in responding to an incident, 4. the President directs the Secretary to assume responsibility for managing the domestic incident.

HSPD-5 also recognizes the role that state, tribal, and local governments; nongovernmental organizations; and the private sector play in managing incidents.

Initial responsibility for managing domestic incidents generally falls on state and local authorities. When their resources are overwhelmed, or when federal property is involved, the federal government provides assistance.

In order to provide a consistent, coordinated, nation-wide approach for emergency operations across all levels of government, HSPD-5 directed DHS to develop and administer a National Incident Management System (NIMS) and a National Response Framework. Together, NIMS and the NRF provide an approach for federal, state, and local governments to effectively prepare for, respond to, and recover from domestic incidents, regardless of cause, size, or complexity.

Critical Infrastructure Identification, Prioritization, and Protection - HSPD-7

HSPD-7 establishes a national policy for federal departments and agencies to identify and prioritize critical U.S. infrastructure and key resources and to protect them from terrorist attacks. Federal departments and agencies will work with state and local governments and the private sector to accomplish this objective. HSPD-7 also identifies Sector-Specific Agencies which, under DHS' overall coordination, lead efforts to protect specific critical sectors and key resources.

In addition, HSPD-7 requires DHS to develop a comprehensive, integrated National Plan for Critical Infrastructure and Key Resources Protection.

Sector-Specific Agencies

Sector-Specific Agencies are agencies responsible for ensuring the protection of a particular resource or part of the national infrastructure. EPA is designated as the Sector-Specific Agency for drinking water and water treatment systems. In addition to being responsible for collaborating across all levels of government, assessing vulnerabilities, and promoting the use of risk-management strategies, EPA must: 1. work with the water sector to reduce the consequences of catastrophic failures not caused by terrorism, 2. collaborate with the private sector to continue the development of information sharing and analysis mechanisms, 3. report to DHS on the Agency's efforts to identify, prioritize, and coordinate the protection of critical infrastructure and key resources.

National Preparedness - PPD-8

National Preparedness Goal

PPD-8 calls for development and maintenance of a National Preparedness Goal defining the core capabilities necessary to prepare for the specific types of incidents posing the greatest risk to the security of the United States. The Goal will establish concrete, measurable, prioritized objectives to mitigate specific threats and vulnerabilities – including regional variations of risk – and emphasize actions intended to achieve an integrated, layered, accessible and all-of-Nation/whole community preparedness approach while optimizing the use of available resources.

DHS, in coordination with other executive departments and agencies, and in consultation with state, local, tribal and territorial governments, the private and non-profit sectors and the general public, submitted the first edition of the National Preparedness Goal in September 2011. The Goals defines success as:

A secure and resilient Nation with the capabilities required across the whole community to prevent, protect against, mitigate, respond to, and recover from the threats and hazards that pose the greatest risk.

The core capabilities contained in the goal are essential for the execution of each of the five mission areas: Prevention, Protection, Mitigation, Response, and Recovery. To assess both preparedness capacity and gaps, each core capability includes capability targets for which measures will be developed. The Goal will be reviewed regularly to evaluate consistency with applicable policies, evolving conditions and the National Incident Management System.

National Preparedness System

The Directive requires a description of the National Preparedness System – an integrated set of guidance, programs and processes, enabling the Nation to meet the National Preparedness Goal. Designed to guide domestic efforts of all levels of government, the private and nonprofit sectors and the public, the National Preparedness System includes guidance for planning, organization, equipment, training and exercises needed to build and maintain domestic capabilities in support of the National Preparedness Goal.

The System description identifies six components to improve national preparedness for a wide range of threats and hazards, such as acts of terrorism, cyber attacks, pandemics and catastrophic natural disasters. The system description explains how the nation will build on current efforts, many of which are already established in the law and have been in use for many years. These six components include:

❑ Identifying and assessing risks;

❑ Estimating capability requirements;

❑ Building or sustaining capabilities;

❑ Developing and implementing plans to deliver those capabilities;

❑ Validating and monitoring progress made towards achieving the National Preparedness Goal; and

❑ Reviewing and updating efforts to promote continuous improvement.

The System will include a series of integrated national planning frameworks covering prevention, protection, mitigation, response and recovery and be built upon scalable, flexible and adaptable coordinating structures. These frameworks are intended to align key roles and responsibilities to deliver capabilities and provide a unified, integrated, accessible system with common terminology. The National Preparedness System includes interagency and departmental operational plans that support each national planning framework with corresponding planning guidance for state, local, tribal and territorial governments.

Other key aspects of the National Preparedness System described in PPD-8 include:

❑ Resource guidance, including arrangements enabling the ability to share personnel;

❑ Equipment guidance, aimed at nationwide interoperability;

❑ National training and exercise program guidance; and

❑ Recommendations and guidance for businesses, communities, families and individuals.

PPD-8 also calls for a comprehensive approach to assess national preparedness. The approach involves measuring operational readiness against target capability levels identified in the Goal.

Campaign to Build and Sustain Preparedness

DHS is tasked with coordinating a comprehensive campaign to build and sustain preparedness nationwide. The campaign will enhance national resilience, federal financial assistance, federal preparedness efforts, and national research and development through proactive public outreach and community-based and private sector programs. The campaign will also tap into existing preparedness efforts across all levels of government and the private and non-profit sectors for a unified approach.

National Preparedness Report

The first annual National Preparedness Report was delivered to the President in early 2012. The report will be based on progress towards achieving the National Preparedness Goal and will serve as a tool to inform the President's budget annually.

Prepared and delivered by DHS, the report requires close coordination with all executive departments and agencies having a role in national preparedness efforts and substantial input from state, local, tribal and territorial governments as well as the private and non-profit sectors and the general public.

EPA's Role Under PPD-8

EPA will participate in the development and execution of the Interagency Planning Frameworks, response activities, training and exercises and contribute to the National Preparedness Report annually.

Clean Water Act and NCP Area Plan Requirements

Requirement	CWA	NCP
When implemented in conjunction with the NCP, be adequate to remove a worst-case discharge, and to mitigate or prevent a substantial threat of such a discharge, from a vessel, offshore facility, or onshore facility operating in or near the area.	✓	✓
A description of the area covered by the plan, including the areas of special economic or environmental importance that might be damaged by a discharge.	✓	✓
A detailed description of the responsibilities of an owner or operator and of federal, state, and local agencies in removing a discharge, and in mitigating or preventing a substantial threat of a discharge.	✓	✓

Requirement	CWA	NCP
A list of equipment (including firefighting equipment), dispersants, or other mitigating substances and devices, and personnel available to an owner or operator and federal, state, and local agencies, to ensure an effective and immediate removal of a discharge, and to ensure mitigation or prevention of a substantial threat of a discharge (this may be provided in an appendix or by reference to other relevant emergency plans (e.g., state or LEPC plans), which may include such equipment lists).	✓	✓
A description of procedures to be followed for obtaining an expedited decision regarding the use of dispersants.	✓	✓
A detailed description of how the plan is integrated into other ACPs and tank vessel, offshore facility, and onshore facility response plans approved by the President, and into operating procedures of the NSFCC.	✓	✓
Compile a list of local scientists, both inside and outside the federal government, with expertise in the environmental effects of spills of the types of oil typically transported in the area, who may be contacted to provide information or, where appropriate, participate in meetings of the scientific support team convened in response to a spill, and describe the procedures to be followed for obtaining an expedited decision regarding the use of dispersants.	✓	
A detailed annex containing a Fish and Wildlife and Sensitive Environments Plan that is consistent with the RCP and NCP. The annex will be prepared in consultation with the U.S. Fish and Wildlife Service, the National Oceanic and Atmospheric Administration (NOAA), and other interested natural resource management agencies. The annex will provide the necessary information and procedures to immediately and effectively respond to discharges that may adversely affect fish and wildlife and their habitat and sensitive environments, including provisions for a response to a worst-case discharge.		✓

This page intentionally left blank

Appendix B:
Area Committees

1. Example AC membership list

    a. *http://www.pwcgov.org/government/dept/FR/Pages/Local-Emergency-Planning-Committee-(LEPC).aspx*

2. Example AC documents

    a. U.S. EPA National Inland Area Contingency Planning Workgroup Charter (January 2010) outlines the vision, mission, objectives, goals, organization, and function of EPA's Inland Area Contingency Planning Workgroup. (Document posted at *http://www.epaosc.org/site/site_profile.aspx?site_id=3857*)

    b. U.S. EPA Inland Area Contingency Planning Brochure (document posted at *http://www.epaosc.org/site/site_profile.aspx?site_id=3857*)

3. Example AC web sites

    a. These Regional Response Team (RRT) web sites include Regional and Area Contingency Plans and other useful planning information:

       - RRT-1 (Maine, Vermont, New Hampshire, Massachusetts, Rhode Island and Connecticut): *http://www.rrt1.nrt.org/production/NRT/RRT1.nsf/AllPages/rrt1Plans.html*

       - RRT-4 (Alabama, Georgia, Florida, Kentucky, Mississippi, North Carolina, South Carolina, Tennessee): *http://www.nrt.org/production/NRT/RRTHome.nsf/Allpages/newrrt_iv-opsmanual.htm*

       - RRT-5 (Illinois, Indiana, Michigan, Minnesota, Ohio, Wisconsin): *http://www.rrt5.org/SubAreas.aspx*

       - RRT-9 (Arizona, California, Nevada): *http://www.rrt9.org/go/site/2763/*

       - RRT-10 (Idaho, Oregon, Washington): *http://www.rrt10nwac.com/NWACP/Default.aspx*

    b. EPA Region 8 (Colorado, Montana, North Dakota, South Dakota, Utah, Wyoming) ACP: *http://www2.epa.gov/region8/area-contingency-plan-acp*

This page intentionally left blank

Appendix C:
ACP Formats, Scope and Organization

1. EPA ACP Model

   a. Model Area Contingency Plan, Volumes I & II (March 1993): Document posted to *http://www.epaosc.org/site/site_profile.aspx?site_id=3857*

2. EPA-only ACPs

   a. EPA Region 1 (Connecticut, Maine, Massachusetts, New Hampshire, Rhode Island, Vermont) ACP: *http://www.epa.gov/region1/er/iacp/index.html*

   b. EPA Region 3 (Delaware, District of Columbia, Maryland, Pennsylvania, West Virginia, Virginia): Draft document posted to *http://www.epaosc.org/site/site_profile.aspx?site_id=3857*

   c. EPA Region 6 (Arkansas, Louisiana, New Mexico, Oklahoma, Texas)ACP: *http://nepis.epa.gov/Exe/ZyPURL.cgi?Dockey=500012JS.txt*

   d. EPA Region 7 (Iowa, Kansas, Missouri, Nebraska): ACP: *http://www.epa.gov/region7/cleanup/superfund/integrated_plan.htm*

3. EPA-USCG joint ACP-related

   a. Region 5: *http://www.rrt5.org/RCPACPMain.aspx*

   b. Region 9: *http://www.rrt9.org/external/content/document/2763/495643/1/RRT-IX%20Regional%20Contingency%20Plan.pdf*

   c. Region 10 ACP: *http://www.rrt10nwac.com/NWACP/Default.aspx*

4. Sub-area plans

   a. EPA Region 5 sub-area plans:

   - Minneapolis-St. Paul: *http://www.umrba.org/hazspills/twincitiesplan.pdf*

   - Northern Michigan: *http://www.great-lakes.net/partners/epa/northmi/*

   - Quad Cities: *http://www.umrba.org/hazspills/quadcitiesplan.pdf*

   - Upper Mississippi: *http://www.umrba.org/hazspills/umrplan.pdf*

   - Peoria, Illinois: *http://www.umrba.org/hazspills/peoriaplan.pdf*

   b. Alaska Sub-area plans: *http://www.akrrt.org/plans.shtml*

   c. Open and Restricted sub-area plan example (Omaha-Council Bluffs)

   (Documents posted at *http://www.epaosc.org/site/site_profile.aspx?site_id=3857*)

5. Geographic response plans

   a. Lower and Middle Columbia River: *http://www.ecy.wa.gov/programs/spills/preparedness/GRP/ColumbiaRiver/ColumbiaRiver.htm*

b. Recommendations for Geographic Response Plan (GRP) Approaches (Document posted at *http://www.epaosc.org/site/site_profile.aspx?site_id=3857*)

6. International contingency plan

   a. Mexico-United States Joint Contingency Plan: *http://www.epa.gov/oem/docs/chem/ipmjcp-e.pdf*

   b. Canada-United States Joint Inland Pollution Contingency Plan: *http://www.epa.gov/oem/docs/er/us_can_jcp_eng.pdf*

   Canada-U.S. Joint Inland Pollution Contingency Plan Regional Annexes and the geographic areas they cover are listed below:

   - Annex I - CANUSWEST (1998) – includes the combined border of the Yukon Territory and British Columbia with U.S. EPA Regions 8 and 10 (Washington, Idaho, Montana, and Alaska): *http://www.epa.gov/osweroe1/content/canada_border.html*

   - Annex II - CANUSPLAIN (2001) – includes the combined border of Alberta, Saskatchewan, and Manitoba with U.S. EPA Regions 5 and 8 (Minnesota, Montana, and North Dakota): *http://www.epa.gov/osweroe1/content/canada_border.html*

   - Annex III - CANUSCENT (2001) – includes the border of Ontario with U.S. EPA Regions 2 and 5 (New York and Minnesota): *http://www.epa.gov/osweroe1/content/canada_border.html*

   - Annex IV - CANUSQUE – includes the inland boundary of Quebec with U.S. EPA Regions 1 and 2 (New Hampshire, Vermont, Maine, and New York): *http://www.epa.gov/osweroe1/content/canada_border.html*

   - Annex V - CANUSEAST (2005) – includes the inland boundary of New Brunswick with U.S. EPA Region 1 (Maine): *http://www.epa.gov/osweroe1/content/canada_border.html*

7. USCG ACP references

   a. Commandant Instruction 16471.3: Area Contingency Plan Organization, Content, Revision Cycle, and Distribution (August 2000): *http://www.uscg.mil/directives/ci/16000-16999/CI_16471_3.pdf*

   b. ACP Development Memo (February 2005) (Document posted at *http://www.epaosc.org/site/site_profile.aspx?site_id=3857*)

Appendix D:
Tools

1. Mapping tools

   a. Minimum Essential Elements for GIS: Document posted to *http://www.epaosc.org/site/site_profile.aspx?site_id=3857*

   b. Example map formats

      - EPA Region 1: *http://www.epa.gov/region1/er/iacp/maps.html*

      - EPA Region 10 jurisdictional boundary tool: *http://gis1.ene.com/epar10/*

   c. Software:

      - LandView® 6: *http://www.census.gov/geo/landview/*

   d. Web resources

      - EPA OSC support site: *http://www.epaosc.org/main/maps.aspx*

      - Compendium of e-mapping applications: *http://www.ehssoftserve.com/geo_mapsinfo.htm* (registration required)

      - NOAA ERMA fact sheet: *http://archive.orr.noaa.gov/book_shelf/1869_ORR-ERMA-07-11.pdf*

      - NOAA ERMA: *http://response.restoration.noaa.gov/erma*

2. Example Sensitive Area inventories

   a. EPA Region 5 Fish and Wildlife annex: *http://www.rrt5.org/RCPACPReferences/FishWildlifeAnnex.aspx*

3. Hazard assessment examples

   a. Natural disaster-related hazards

      - Region 6 Natural Disaster Workgroup: *http://www.epaosc.org/site/site_profile.aspx?site_id=4907*

4. Historical preservation resources

   a. National Park Service list of historic preservation officers: *http://www.nps.gov/nr/shpolist.htm*

5. Riverine spill modeling

   a. Ohio River Valley Water Sanitation Commission (ORSANCO)modeling tool: *http://www.orsanco.org/emergency-response-program*

This page intentionally left blank

Appendix E:
Contact/Notification Lists

1. Example ACP contact list (Document posted to: *http://www.epaosc.org/site/site_profile.aspx?site_id=3857*)

2. Example Incident Notification lists

    a. Omaha-Council Bluffs notification list (Document posted to: *http://www.epaosc.org/site/site_profile.aspx?site_id=3857*)

    b. Truckee River incident notification list (Document posted to: *http://www.epaosc.org/site/site_profile.aspx?site_id=3857*)

This page intentionally left blank

Appendix F:
Resource Inventory Development

1. Example response resource inventories

    a. EPA Region 10 equipment inventory: http://www.rrt10nwac.com/Equipment.aspx

    b. EPA Environmental Response Team equipment inventory information:

    - Two Turner C7 Fluorometers, including one 50 Meter Cable: Contains multi sensor array, but primarily used for in-water detection of crude or refined oil, could be integrated into existing vessel CTD platform. http://www.turnerdesigns.com/products/submersible-fluorometer/cyclops-7-submersible-fluorescence-and-turbidity-sensors

    - Turner 10AU Flow Thru Fluorometer: "Old Reliable" Model, has a pump and internal lamp, bench top application, can run individual samples or pump/flow thru continuous sampling, can use for dye tracer studies also, could use to monitor DW intakes or other water intakes for oil contamination. http://www.turnerdesigns.com/t2/doc/manuals/10au_manual.pdf

    - Site Lab UV3100 from CyberSense: Recently acquired, for soil and water samples, simple extraction required, bench top portable application, applies UV fluorescence to extract yielding numeric concentration, can analyze oil fractions for TPH, GRO, DRO, and PCBs, purchase standards for calibration. http://www.cysense.com/images/upload/docum/CTPN200518_UVF3100_PAH%20TPH%20and%20PCB.pdf

    - PhotoVac Voyager Portable GC: Analysis for VOCs, Chlorinated solvents, benchtop and potentially used over the shoulder for compounds in air, can use heating element to purge volatiles in water and analyze headspace, good sensitivity, mostly spills or site characterization applications, also soil gas. http://www.equipcoservices.com/pdf/manuals/photovac_voyager.pdf

This page intentionally left blank

Appendix G:
Response Strategy Development

1. Oil spill counter-measure examples

   a. Dispersant authorization

   Subpart J of the National Oil and Hazardous Substances Pollution Contingency Plan (NCP) directs EPA to prepare a schedule of dispersants, other chemicals, and oil spill mitigating devices and substances that may be used to remove or control oil discharges.

   - NCP Subpart J: Use of Dispersants and Other Chemicals - 40 CFR 300.900 - 300.920 (current rule): http://www.epa.gov/emergencies/docs/oil/cfr/900_920.pdf

   - September 15, 1994, National Oil and Hazardous Substances Pollution Contingency Plan; Final Rule. 59 FR 47384 (current regulations): http://www.gpo.gov/fdsys/pkg/FR-1994-09-15/html/94-22347.htm

   b. Dispersant pre-approval examples

   - RRT-6 FOSC Dispersant Pre-approval Guidelines and Checklist: http://www.losco.state.la.us/pdf_docs/RRT6_Dispersant_Preapproval_2001.pdf

   - Use of Dispersants in Region IV: http://www.nrt.org/production/NRT/RRTHome.nsf/Resources/DUP/$file/1-RRT4DISP.PDF

   c. Cleaning agents

   - http://nepis.epa.gov/Exe/ZyPURL.cgi?Dockey=30002UZK.txt

   d. Bioremediation

   - epaosc.net: http://www.epaosc.org/site/doc_list.aspx?site_id=ERTREAC016

   - Literature Review on the Use of Commercial Bioremediation Agents for Cleanup of Oil-Contaminated Estuarine Environments: http://www.epa.gov/osweroe1/docs/oil/edu/litreviewbiormd.pdf

   - Guidelines for the Bioremediation of Oil-Contaminated Salt Marshes: http://www.epa.gov/osweroe1/docs/oil/edu/saltmarshbiormd.pdf

   - Guidelines for the Bioremediation of Marine Shorelines and Freshwater Wetlands: http://www.epa.gov/osweroe1/docs/oil/edu/bioremed.pdf

   - National Response Team Fact Sheet on Bioremediation Technologies: http://www.epa.gov/osweroe1/docs/oil/edu/biofact.pdf

   e. In-Situ Burning

   - Inland In-Situ Burning of Oil Spills: Regulations and Authorizations: http://www.epa.gov/oem/docs/oil/fss/fss09/dehaven.pdf

   - NOAA guidance for monitoring in-situ burning operations: http://response.restoration.noaa.gov/ISB

f. Dispersant Monitoring System

- NOAA Special Monitoring of Applied Response Technologies (SMART): http://response.restoration.noaa.gov/smart

2. FRP-related response strategies

a. Example oil spill tactical response plan: (Document posted at: http://www.epaosc.org/site/site_profile.aspx?site_id=3857)

3. Sensitive resource-related response strategies

a. Examples can be found at:

- http://www.epa.gov/osweroe1/content/fss/00table.htm

- http://www.epa.gov/swercepp/web/content/fss/09table.htm

4. National Response Team Subsea and Surface Dispersant Guidance

a. The guidance is currently under development/review.

5. Memo from Mathy Stanislaus, OSWER Assistant Administrator, dated Nov. 2, 2010, to the Regions requesting updates to the ACPs.

a. Link to the memo can be found at: http://www.epaosc.org/site/site_profile.aspx?site_id=3857

6. Mechanical cleanup technologies

a. Examples can be found at:

- http://www.epa.gov/oem/docs/oil/edu/oilspill_book/chap2.pdf

- http://www.epa.gov/oem/content/learning/oiltech.htm

Appendix H: Resources for Assistance in ACP Development

1. EPA Regional and Headquarters offices contact information (contacts for ACP matters):
   http://www.epa.gov/oem/content/regional.htm

This page intentionally left blank

Appendix I:
Volunteers

The National Response Team (NRT) has released the NRT Use of Volunteers Guidelines for Oil Spills. The Guidelines can be found on the NRT Website (*www.nrt.org*; Guidance, Technical Assistance & Planning; Use of Volunteers Guidelines for Oil Spills) or by clicking this link: *http://www.nrt.org/production/NRT/NRTWeb.nsf/AllAttachmentsByTitle/SA-1080NRT_Use_ of_Volunteers_Guidelines_for_Oil_Spills_FINAL_signatures_inserted_Version_28-Sept-2012. pdf/$File/NRT_Use_of_Volunteers_Guidelines_for_Oil_Spills_FINAL_signatures_inserted_ Version_28-Sept-2012.pdf?OpenElement*

This page intentionally left blank

Appendix J: Acronyms

ACP	Area Contingency Plan
CAMEO	Computer-Aided Management of Emergency Operations
CERCLA	Comprehensive Environmental Response, Compensation, and Liability Act
CERCLIS	Comprehensive Environmental Response, Compensation, and Liability Information System
COTP	Captains of the Port
CWA	Clean Water Act
DHS	Department of Homeland Security
DoD	Department of Defense
DOT	Department of Transportation
DRG	District Response Group (USCG)
EMAC	Emergency Management Assistance Compact
EPA	Environmental Protection Agency
EPCRA	Emergency Planning and Community Right-to-Know Act
ERMA	Environmental Response Management Application (NOAA)
ESA	Endangered Species Act
ESF	Emergency Support Function
FEMA	Federal Emergency Management Agency
FHWA	Federal Highway Administration (DOT)
FMCSA	Federal Motor Carrier Safety Administration (DOT)

FOSC	Federal On-Scene Coordinator
FRA	Federal Railroad Administration (DOT)
FRP	Facility Response Plan
FRS	Facility Registry System
GIS	Geographic Information System
HHS	Department of Health and Human Services
HSPD	Homeland Security Presidential Directive
IC	Incident Commander
ICS	Incident Command System
JIC	Joint Information Center
LEPC	Local Emergency Planning Committee
MCP	Mobile Command Post
NCP	National Oil and Hazardous Substances Pollution Contingency Plan
NEBA	Net Environmental Benefits Analysis
NGO	Non-Governmental Organization
NIMS	National Incident Management System
NOAA	National Oceanic and Atmospheric Administration
NRF	National Response Framework
NRS	National Response System
NSFCC	National Strike Force Coordination Center (USCG)
OEI	Office of Environmental Information (OEI)
OEM	Office of Emergency Management (EPA)

OPA 90	Oil Pollution Act of 1990
OSC	On-Scene Coordinator
PHMSA	Pipeline and Hazardous Materials Safety Administration (DOT)
PPD	Presidential Policy Directive
PPE	Personal Protective Equipment
RCP	Regional Contingency Plan
RMP	Risk Management Plan
RRT	Regional Response Team
SERC	State Emergency Response Commission
SSC	Scientific Support Coordinator
TRI	Toxics Release Inventory
UC	Unified Command
USCG	United States Coast Guard
USDA	United States Department of Agriculture
USGS	U.S. Geological Survey
VRP	Vessel Response Plan